Tewendé Lionel Kevin BÉRÉ

Influence of 2 sowing dates on agro-morphological parameters

AF671789

Tewendé Lionel Kevin BÉRÉ

Influence of 2 sowing dates on agro-morphological parameters

of sweet-stemmed sorghum genotypes

ScienciaScripts

Imprint

Any brand names and product names mentioned in this book are subject to trademark, brand or patent protection and are trademarks or registered trademarks of their respective holders. The use of brand names, product names, common names, trade names, product descriptions etc. even without a particular marking in this work is in no way to be construed to mean that such names may be regarded as unrestricted in respect of trademark and brand protection legislation and could thus be used by anyone.

Cover image: www.ingimage.com

This book is a translation from the original published under ISBN 978-620-6-72400-1.

Publisher:
Sciencia Scripts
is a trademark of
Dodo Books Indian Ocean Ltd. and OmniScriptum S.R.L publishing group

120 High Road, East Finchley, London, N2 9ED, United Kingdom
Str. Armeneasca 28/1, office 1, Chisinau MD-2012, Republic of Moldova, Europe
Printed at: see last page
ISBN: 978-620-8-23709-7

Copyright © Tewendé Lionel Kevin BÉRÉ
Copyright © 2024 Dodo Books Indian Ocean Ltd. and OmniScriptum S.R.L publishing group

Contents

DEDICATION 2
ACKNOWLEDGEMENTS 3
SUMMARY 6
INTRODUCTION 8
CHAPTER I 11
CHAPTER II 18
CHAPTER III 27
CONCLUSION AND OUTLOOK 43
BIBLIOGRAPHY 45
APPENDIX 50

DEDICATION

Ppapa: Mr BERE Francois

Mother: Mrs BERE/KABRE T J. R Viviane.

ACKNOWLEDGEMENTS

Au terme de ce тётоие, nos remerciements vont a 1 endroit de tous ceux qui, de pres ou de loin, ont particip6 a son ёlaЬoraPoп. Nous tenons particulierement a dire merci a :

- Mr SAWADOGO Mahamadou, Professeur Titulaire a i'universile JOSEPH KI- ZERBO, actuel Secretaire дёиёral du ministere des enseignements supërieurs, de la recherche scientifique et de l'innovation et Directeur du Laboratoire Biosciences, for having accepted us in this training course and for agreeing to chair the defence jury;
- Mrs BATIONO/KANDO Pauline, Professeur Titulaire a i'universile JOSEPH KI-ZERBO, for the efforts made as our teacher and Director of the present mëmoire. Your advice, criticism and suggestions have been very helpful. Thank you for agreeing to participate in the evaluation of this work as a member of the defence jury;
- Mr SAWADOGO Nerbëwendë, Maitre de Confërences a l'UniversE Joseph KI-ZERBO, Coordinator of the master ; for the advice as our teacher and co-supervisor of this work. Thank you for agreeing to co-direct this dissertation, for appreciating the form and content of this document and for participating in the defence jury;
- Mr TRAORE Renan Ernest, Maitre de confërence a l'universE JOSEPH KI- ZERBO for agreeing to examine this document as a member of the jury and to correct it in form and content;
- All members of the jury ;
- Mr SAWADOGO Boureima, Assistant at the University of Fada N'Gourma, for

We would like to take this opportunity to express our sincere gratitude for their help in setting up the trial, entering the data and correcting the form and content of the report;

- Mr TIENDREBEOGO K. Fidele, doctor in gënëtique and amëlioration of plants for having prëtë forte in the correction of this mëmoire. Your criticisms and suggestions have ëtë bënëfiques us ;
- Mr BERE Andre, retired veterinary doctor, and Mr BORO Oumar, doctoral student in veterinary medicine.

Gënëtique et Agronomie, for agreeing to correct this mëmoire. Your criticisms and suggestions have ëtë been useful to us;

- I would like to thank all my teachers for the learning and advice they gave me, and the data analysis methods they used during their lessons. Thank you to Messrs NANEMA K. Romaric, Maitre de confërences ; KIEBRE Zakaria, Maitre-assistant et OUEDRAOGO Mahamadi Hamed, Maitre-assistant ; tous a i'universite JOSEPH KI- ZERBO ;
- All my friends, comrades, cousins, brothers and sisters, Mr SOUMBOUGMA Benoit, doctoral student in gënëtique et amëlioration des plantes ;
- Monsieur NIKIEMA Oumarou, Technicien de la station de recherche de Gampëla ;

TONDE W. Hermann, student in professional Master SVRPG ; BAMOUNI Ingrid Dieudonne, doctor in mëdecine ; BERE Bernard N, student in English ; BOUTIANA Emmanuel, student in English ; GUIATIN Gaetan, student in economics ; BERE Patrick Junior ; Mademoiselle OUEDA Lucie ; ILBOUDO K. Osee; for their respective mutiform support from the setting up of the trials, through data collection, to analysis and interpretation.

- All those who, directly or indirectly, have participated in the success of this project, are hereby congratulated.

of this document and whose names could not be ascertained.

SUMMARY

Sweet stem sorghum [*Sorghum bicolor (L.)* Moench] is a plant traditionally grown around concessions in Burkina Faso. Despite its low exploitation in Burkina Faso, sweet stem sorghum is full of potential, especially due to its high accumulation of sugars in its stems and the quality of its fodder. The aim of this study was to determine a favourable sowing period for optimum expression of the agromorphological characteristics of sweet-stem sorghum and their response to photoperiod variation. Thus, 29 genotypes from the gene bank of the Biosciences Laboratory of the JOSEPH KI- ZERBO University were evaluated using 27 variables, including 07 qualitative variables and 20 quantitative variables in a Fisher block design with three repetitions. This study showed that all 07 qualitative variables and 02 quantitative variables, in particular peduncle length and internode length (LPE, LOE), were not influenced by sowing date. However, 18 quantitative variables were influenced by sowing date. This influence resulted in a significant drop in the average performance of 13 quantitative variables during late sowing, in particular characteristics relating to the cycle (NJG, NJE, NJF), characteristics related to leaf dimensions (LOF, LAF), characteristics related to plant dimensions (NEN, HPL, DTP), characteristics related to panicle dimensions (LOP, LAP, PPA), sugar content at the milky stage (BSL) and aerial tillering (TLA). We noted an increase in performance for 05 quantitative variables, namely vegetative and useful tillering (TLV, TLU), number of days to emergence (NJL) and sugar content at the pale and hard grain stages (BSP and BGD). Five (05) genotypes (namely 124, 108, 02, 79 and 109) were not very susceptible while 11 genotypes were highly susceptible. Genotype 85 was the most photoperiodic (with Kp = 0.861) and 124 the least photoperiodic (with Kp = 0.125). The genotypes had a higher brix at the second sowing date than at the first sowing date. These results could be used in the sorghum sugar improvement programme.

Keywords: *Sorghum bicolor,* agro-morphological variability, photoperiodism, sowing, Burkina Faso.

Cëréales cover 4.2 million hectares, i.e. three quarters of the cultivated area in Burkina Faso (Ministere de 1 agriculture et de la securite alimentaire-Situation de reference des principales filieres agricoles au Burkina Faso, April 2013). From 2009 to 2018, the area farmed in Burkina Faso increased by 1519364 ha, or 32%. Over the period from 2009 to 2018, farmland remained mainly devoted to cereals. Indeed, cereals accounted for 80% of the total area over the period, followed by cash crops (24%) and other food crops (5%) (EPA/DGESS/MAAH, 2018). Sorghum is the most widely grown cereal in Burkina Faso. It is currently grown on an estimated 1.3 million hectares. Sorghum is grown throughout Burkina Faso during the rainy season (Maisonneuve et Larose, 1991). Sorghum is one of the most widely consumed cereals in Burkina Faso. Sorghum's share of total calorie requirements is estimated at 17.5% on average over the period 2005-2011, rising from 16.7% to 19% (FAOSTAT, 2014). According to the ECOFIN Agency (2018), in Burkina Faso, the results of the 2017/2018 agricultural season published by the government show a cereal harvest of 4 million tonnes. Different types of sorghum are grown in Burkina Faso, including sweet stalk sorghum. Sweet-stem sorghum is traditionally grown around concessions, mainly in association with mai's, common sorghum or other NEBIE crops (2014). According to NEBIE (2009), the low production of this sorghum is due to a lack of knowledge of its potential. However, sweet stalk sorghum is well known in other countries, notably China, the USA and many others, and is used as a raw material for the manufacture of consumables such as syrup, sugar and biofuel ethanol (GUIYING *et al.*, 2000; FAO 2002; ICRISAT, 2004). It is also used for fodder, and the sugars accumulated in high concentrations (Brix) in the stems of this type of sorghum are used as an energy or food supplement by local populations (NEBIE *et al.* 2013). Most of the studies carried out in Burkina Faso on sweet stem sorghum have focused on its genetic diversity (NEBIE, 2014) and its potential for sugar production, particularly Brix (NEBIE *et al.* ,2013). In a context of climate change characterised by the late onset and shortening of the rainy season, as well as irregular rainfall, it appears necessary to develop more resilient varieties. Thus, the present study entitled "influence de deux dates de semis sur les parametres agro-morphologiques de genotypes de sorgho a tige sucree" (influence of two sowing dates on the agro-morphological parameters of sweet stem sorghum genotypes) was initiated to determine the impact of a sowing offset on the agro-morphological variability of 29 sweet stem sorghum genotypes. Specifically, the aim was (i) to assess the impact of each sowing date on the expression of agro-morphological variables of sweet stem sorghum genotypes; (ii) to determine the impact of 'delayed' / 'advanced' sowing on the agro-morphological variability of sweet stem sorghum genotypes; (iii) to assess the degree of sensitivity to photoperiod variation.

The present document is structured around three chapters with a general introduction and conclusion. The first chapter is devoted to a literature review on sweet stem sorghum, the second chapter deals with the materials and methods used and the third chapter is devoted to the results obtained and their discussion.

CHAPTER I

GENERAL INFORMATION ON SWEET STEM SORGHUM

I. History

Sorghum is one of the oldest cultivated plants in the world (OUEDRAOGO, 2014). Domesticated around 8000 years ago, sorghum originated in East Africa (Ethiopia, Sudan). It is particularly well adapted to semi-arid tropical regions where it remains, along with millet, an essential food crop for the food security of hundreds of millions of people in the dry regions of sub-Saharan Africa and South Asia (EPSO, 2015).

DOGGETT (1965) reports archaeological evidence suggesting that the practice of domesticating cereals was introduced into Egypt from Ethiopia around 3000 BC. According to him, it is possible that the domestication of sorghum began at this time. Sweet stalk sorghum is thought to have been introduced to America at the beginning of the 19th century[e] . It had sweet stalks from which juice was extracted to make sugar and syrup (HARLAN and DE WET, 1972).

II. Taxonomy

Sorghum [*Sorghum bicolor* L. (Moench)] is an annual herbaceous plant in the family *Poaceae* (ex-Graminees), subfamily *Panico'ideae,* tribe *Andropogoneae* and genus *Sorghum* (DOGGET, 1988). During the 16th century, sorghum was first designated under different names, namely *Milliumsaracenaceum, Milliumindicumsivemelica, Milliumindicum and Milliumaethiopicum*. Modern taxonomy only took up the name from Linnaeus, who was the first to describe sorghum under the name *Holcus*, and described 07 species, of which 03 are still part of the genus *Sorghum*: *Holcussaccaratus*, *Holcussorghum* and *Holcusbicolor*. However, the current systëmatique is inspired by the bases given by Moench, who was the first to define the genus Sorghum and the species *Sorghum bicolor* (L) Moench (OUEDRAOGO, 2014).

Figure I is a schematic representation of the classification of sorghum within the *Poaceae* family.

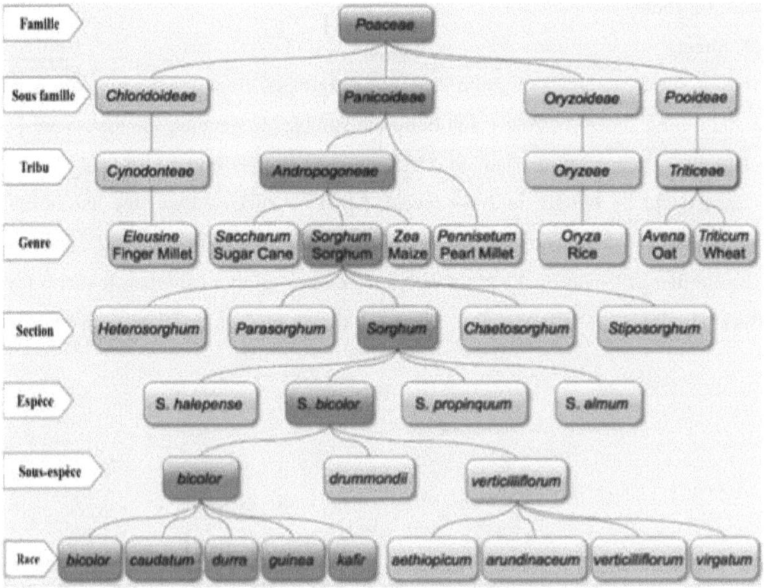

Figure I : Schematic representation of the classification of sorghum within the Poaceae family (after Bouchet, 2008)

III. Biology

III.1. 1. Morphology of the vegetative apparatus

III.1.1.1. Root

Like other types of cultivated sorghum, sugar sorghums are characterised by a powerful root system, which largely explains their ability to withstand severe water stress. The roots are not very deep in the soil, but are vigorous (GRASSI, 2001; NYABYENDA, 2005). The sorghum root system is well developed, with numerous root hairs (almost twice as many as mat's, for example) (HOUSE, 1987). According to NEBIE (2014), roots explore the soil in all directions and originate from the lower internodes.

III.1.2.2. Chaume

Sweet stalk sorghum or 'kao- liang' (Chinese millet) varies in height from 1.20 to 4 m and can reach 5 m, depending on the variëtës and growing conditions. Morphologically, it is similar to other cultivated sorghums (FAO, 2002; GNANSOUNOU *et al.*, 2004; ICRISAT, 2006 b). The stem is solid with a hard cortex or bark and a softer pith. The pith is sweet and juicy. Sorghum is a C4 type plant. According to NEBIE (2014), this photosynthesis mechanism enables it to store large quantities of sucrose. The reserve organ par excellence is the stalk in sweet stalk sorghum.

III.1.3.3. Sheets

The leaves of sweet-stemmed sorghum are simple, hairless, erect or drooping, alternate, with a ligule and a long sheath narrowly adjoining the stubble. In general, the leaves are between 30 cm and 135 cm long and 6 cm to 13 cm wide (SINGH and LOHITHASWA, 2006). Studies carried out by NEBIE in 2009 revealed that in Burkina Faso, the leaves of sweet-stemmed sorghum in the Centre-North region are 61 cm to 85 cm long and 5 cm to 11 cm wide. The number of leaves on the plant varies from one variety to another. It varies from 6 to 14 in Burkina cultivars (NEBIE 2009). Figure II shows a sorghum plant with only its main stem.

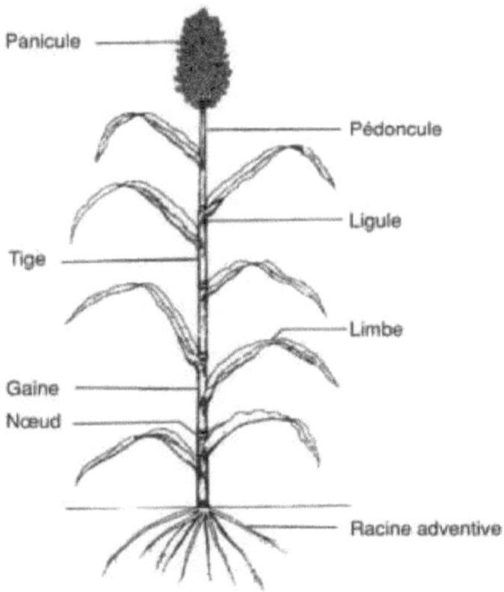

Figure II: Diagram of a sorghum plant with a single main stem (after Clerget, 2004)

III.2. 2. Morphology of the reproductive system

III.2.1.1. Inflorescence

The sorghum inflorescence is a panicle. The sorghum panicle consists of a central axis, the rachis, which carries the branches. These branches in turn support a pair of tpillets, one sessile and fertile, the other ptdicellt and sttrile (CHANTEREAU and NICOU, 1991). Each tepillet bears two (2) flowers, of which only the uppermost is complete. Each complete flower has two (02) stigmas and three (03) tetamines (AHMADI *et al.*, 2002). HOUSE (1987) states that the inflorescence of the tillers can occur simultaneously with that of the main stem or after the latter.

III.2.2.2. Grain

The sorghum grain is a caryopsis composed of three main parts: the envelope which constitutes the ptricarp (6 to 8% of the dry weight), the reserve tissue or albumen also called endosperm (80 to 84%) and the embryo which occupies 7.8 to 12.1% of the dry weight of the grain; ASIEDU (1989). According to HOUSE (1987), the fully matured grain is 3.5 to 5 mm long, 2.5 to 4.5 mm wide, with a weight of 1000 grains varying from 60 to 85 g.

III.3. 3. Sorghum physiology

III.3.1.1. Germination and emergence

Grain germination can take two different forms, hypogte and tpigte. These two forms are distinguished by the position of the cotyltdons in relation to the soil during germination. Germination is hypogynous, also known as cryptocotylar germination, in sorghum as in all grains (DUKE, 1965). Germination is the passage of the grain from a state of slowed life to a state of active life. It is followed by emergence. According to ZOLIKPO (2011), sorghum seedlings take 3 to 10 days to emerge.

III.3.2.2. Trim

The Internet user defines tillering as the ability of gramineous plants to multiply from a single shoot. The ability to till depends as much on the variety as on environmental conditions, in this case population density, nitrogen supply, temperature and photoperiod (DOGGETT, 1988). CHANTEREAU and NICOU (1991) report poor tillering in tropical sorghums of the *guinea* type.

III.3.3.3. Reproduction

In sorghum, flowering begins at the top of the panicle and progresses downwards. Flowering continues down the inflorescence for 4 or 5 days. Each panicle can contain up to 6,000 florets (QUINBY and KARPER, 1947). Flowering begins when the plant is between 50 and 60 cm high and the duration of the day is shorter than that of the night (CHANTEREAU and NICOU, 1991; ZONGO, 1991). Sorghum is mainly self-pollinated (although cross-pollination accounts for about 2-10%) (HOUSE, 1987). In sorghum, pollen collects on the stigma to fertilise the ovule and form a nucleus with 2n chromosomes.

III.4. 4. Photoperiodism

Photoperiodism, an important phenomenon in the animal and plant world, reflects the influence of the length of day and night on various physiological reactions (in this case the flowering of plants). According to FRANQUIN (1971), photoperiodism is the phenomenon whereby light intervenes not by its intensity but by its duration during the 24-hour cycle. Control of the transition between the vegetative and reproductive phases is generally dependent on two environmental signals: day length (or photoperiod) and temperature

(QUINBY *et al.*, 1973). According to OUEDRAOGO (2014), sorghum is generally photoperiodic and this photoperiodism may be more accentuated in some genotypes than others. Sorghum's sensitivity to photoperiod (photoperiodism) then causes a shortening of the cycle when sowing is delayed, favouring flowering grouped with the end of the rainy season (GARNER and ALLARD, 1923a, COCHEME and FRANQUIN, 1967). The later the sowing, the shorter the plants. This reduction in height results directly from a reduction in the number of phytomeres produced, but rarely in their size (BEZOT, 1963; CLERGET *et al.*, 2008b). It should also be noted that the sugar content in the stems depends on the phenology of the plant, dictated by the length of the sowing-flowering cycle and the photoperiod (GUTJAHR, 2012).

III.5. 5. "Stay-green" sorghum

Stay-green" refers to the sorghum plant's ability to keep a good number of green leaves after flowering in order to ensure that the grains are adequately filled with nitrogen compounds and sugar (OUEDRAOGO, 2014). This characteristic has been targeted by Australian breeding programmes, which have led to the development of sorghums that are smaller but more vigorous, and ensure a high-quality harvest.

(GAUFICHON *et al.,* 2010). Stay-green" is a trait that reflects the plant's ability to tolerate post-flowering drought.

IV. Ecology

V. .1. Temperature and Light

According to ZOLIKPO, the optimum temperature for sorghum grain germination is 2735°C. However, too great a drop in temperature (temperatures below 1215°C, for example) would result in sterility during the flowering phase (OUEDRAOGO, 2014).

Sorghum is a short-day plant that reacts to photoperiod in different ways. At high latitudes, some tropical cultivars do not flower or set seed. In the United States, Australia and India, cultivars ranging from average to virtually insensitive to photoperiod have been recorded (OUEDRAOGO, 2014). In the case of sorghum, not all varieties are photoperiodic, but those grown by farmers in Africa are often nyctiperiodic. This adaptive characteristic enables local varieties to synchronise their flowering with the end of the rainy season (CURTIS, 1968; KOURESSY *et al.*, 2008).

VI. 2. Water requirements

Known as the "camel plant" in India, sweet sorghum is highly adaptable, highly resistant to drought and saline-alkaline soils, and tolerates waterlogging well (FAO, 2002). According to SENE (1995), sorghum needs between 400 mm of rainfall for a short cycle and 700 mm for a long cycle. Rainfall must be well distributed over time to ensure maximum yield.

VII.Uses of sugar sorghums

VIII. Industrial uses

Particularly in northern countries, sweet-stemmed sorghums are the most widely grown. The stalks are harvested before the grains ripen, to extract the juice. In a good variety, the juice represents at least 50% of the stalk mass (ASIEDU, 1989). According to SCHAFFER and GOURLEY (1982) and QAZI et *al* (2012), sucrose (non-reducing sugar), glucose and fructose (reducing sugars) are the main sugars found in the stem juice of sweet sorghums, with a high proportion of sucrose of around 80%. However, these proportions vary according to the variëtës. The juice, which has a more or less neutral pH (around 7), according to CHAVAN et *al* (2009), also contains proteins, starch, fibre and calcium.

The sorghum sugar juice extracted in this way can be fermented to produce ethanol. Up to 7,000 litres of ethanol can be produced from one hectare of sweet stem sorghum (FAO, 2002). According to ICRISAT (2006 a), ethanol has interesting ënergëtic properties. In fact, 1.5 litres of juice has the same energy content as a litre of petrol.

The stem juice, seeds and industrial by-products of sugar sorghum are used for a variety of purposes. But the most common uses are the production of syrup and fodder. Sorghum syrup is obtained from certain variëtës called "variëtës syrup". The juice of these sorghums contains very little sucrose, although there is a significant amount of non-crystallisable sugar (SCHAFFER and GOURLEY, 1982; ASIEDU, 1989).

IX. . Non-industrial uses

In developing countries, sweet grain sorghums are grown mainly for their grain, while sweet stalk sorghums are grown for the sugar juice contained in the stalk, or secondarily for fodder. Stems are eaten as a delicacy by chewing to suck the juice contained in the milky or pasty stage (VIGUIER, 1947; HARLAN and DE WET, 1972; NEBIE, 2014). The leaves, stalks and panicles, whether or not seeded, are used for animal feed. In general, sorghum, in addition to fodder, can be шШзё in composting (straw). The stalk can also be used to make mats, as fuel in many homes, etc. (NEBIE, 2009).

CHAPTER II

MATERIALS AND METHODS

I. Experimental site

The trials were carried out at the Institut de Dëveloppement Rural (IDR) experimental station in Gampëla from June to December 2019. This station is located 18 km from Ouagadougou, on the Ouagadougou-Fada N'Gourma axis, at 12°5' north latitude and 1°12' west longitude. Figure III shows the location map of the Gampëla station.

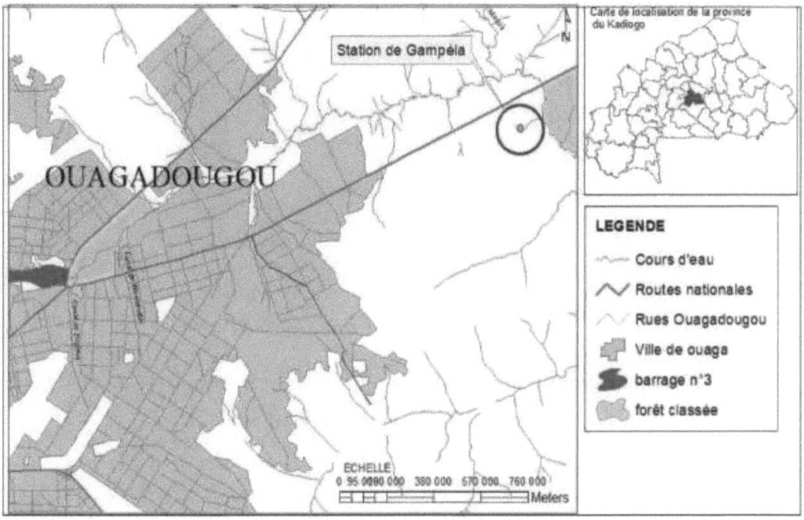

Figure III: Location map of the Gampela station (National Topographic Data Base (NTDB), 2012)

The climate at the station is Sudanian, characterised by a short rainy season from June to October and a long dry season from November to May (THIOMBIANO and KAMPMANN, 2010). Cumulative rainfall during the 2018-2019 agricultural season ëlë 852.7 mm spread over eight (08) months. July ëlë the wettest month with 321.3 mm of rainfall. Temperatures during the same season ranged from 25.6°C to 34.7°C. Figure IV shows the rainfall-temperature curve for the Gampëla station according to the Direction de la Mëtëorologie Nationale, 2019.

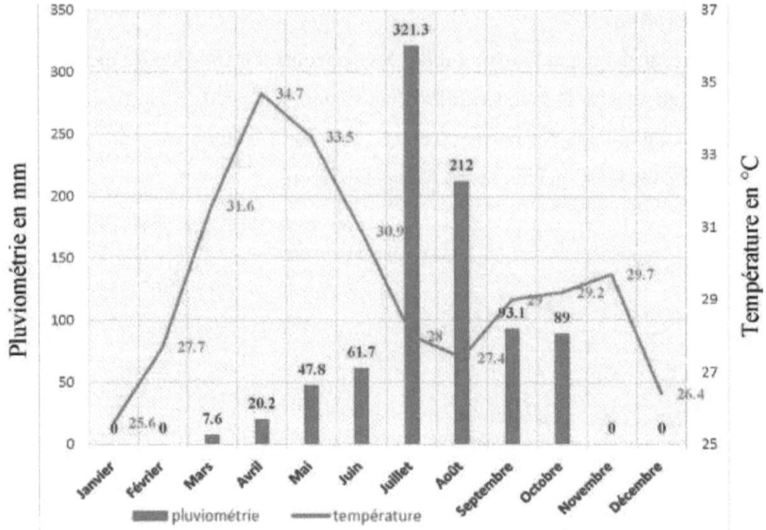

Figure IV: Rainfall-temperature curve for the Gampela station (Direction de la meteorologie nationale, 2019)

II. Plant material

The vëgëtal material is composed of 29 sweet stem sorghum gënotypes. These 29 gënotypes were selected randomly from the germplasm of the Biosciences Laboratory of the Plant Genetics and Improvement Team (EGAP) at the Joseph KI-ZERBO University.

III. Methods

III.1. 1. Experimental set-up

Two trials were set up, the first in June and the second in July. The experimental set-up used for each of the two trials was a Fisher block with 3 rëpëtitions. Each rëpëtition comprises 29 lines with one gënotype per line. Each line has 14 stacks. The row spacing and spacing were 80 cm and 40 cm respectively. Two border lines were placed on either side of each row. The inter-rëpëtition is 2 m. Each device has a total area of approximately 502 m^2 with a length ële of 25.6 m and a width of 19.6 m. Figure V below shows the scheëma of the experimental set-up for each trial.

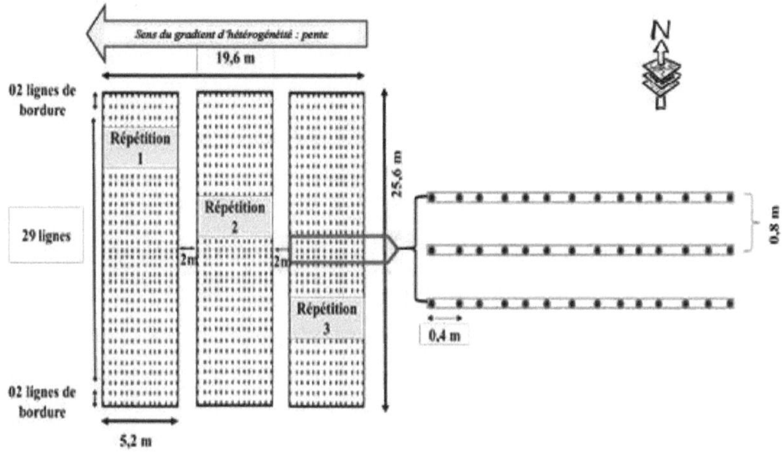

Figure V: Schema of the Fisher block design for each of the two trials (BERE, 2019)

III.2. 2. Cultivation techniques

The expërimentation a соттепсё by a germination effects test on 40 ateatorily selected gënotypes. Twenty-nine (29) gënotypes were ël.ë sëlectionnës for 1'essai on the basis of germination rate. The first oprere sowings on field n°1 occurred on 29 June 2019. The second opërës sowings on field n°2 occurred on 23 July 2019. Prior to trial establishment, both plots had Ьё^й^ё of ploughing and levelling. Two weedings were ё1ё ^ necessary; the first, prëcëdë from a dëmariage to a plant occurred three (03) weeks after the sowing date while the second weeding occurred about seven (07) weeks after the sowing date. A ridging was carried out towards the end of the vëtative development of the plants in order to counteract lodging caused by high winds. Two amendments, the first with NPK and the second with urea, were applied to the two plots at a rate of 100kg/ha.

III.3. 3. Data collection

Two types of data were collected. Quantitative variables and qualitative variables. A total of 27 variables were used, including 20 quantitative variables and seven (07) qualitative variables.

III.3.1.1. Quantitative variables

The quantitative variables were measured directly in the field, with the exception of panicle length (LOP) and width (LAP), panicle weight (PPA) and brix (BRIX). The brix represents the sucrose fraction in a liquid, i.e. the percentage of soluble dry matter. The length of the panicle is measured from the base to the top of the panicle using a graduated tape. The width is measured in the lower third of the panicle. Weight was measured using a digital scale and brix using a portable refractometer. The following variables were measured along the entire line:

- the number of days to emergence (NJL) corresponds to the number of days between sowing and emergence of half the bunches per row;
- the number of days to swelling (NJG) is the number of days between sowing and swelling of 50% of the plants per row;
- the number of days to heading (NJE), which is the number of days between sowing and the heading of 50% of the plants per row;
- the number of days to flowering (NJF) corresponds respectively to the number of days between sowing and flowering of 50% of the main panicles of the ecotype per line.

The quantitative variables below were measured at the ripening stage on 03 plants randomly selected from each row. These were :

- The length (LOF) and width (LAF) of the leaf were measured on the third leaf below the panicle. The choice of the third sub-panicular leaf is justified by the fact that the leaf in question is no longer actively growing and is also less affected by any stress caused by water deficit. The width of the leaf was measured at the widest part of the same leaf;
- of the length of the inter-knot (LOE), which was estimated by measuring the half-distance between the third knot (3eme) and the fifth knot (5eme) from the vertex;
- the length of the peduncle (LPE) measured from the last node to the first branching of the panicles;
- of the length (LOP) and width (LAP) of the panicle corresponding respectively to the distance sëparant the base of the panicle to the top of the panicle and to the distance of the widest part of the panicle. All measurements relating to leaf and panicle dimensions were ële made using a graduated tape;
- plant height (HPL), which is the distance from the top of the main panicle to the base of the plant on the ground. It was measured using a graduated scale.
- and stem diameter (DTI) measured on the first internode of the main stem from the base. This was done using a calliper.

Vegetative tillering (TLV), useful tillering (TLU) and aerial tillering (TLA) were scored by counting, respectively, the number of tillers that developed at the first node, the number of tillers that produced panicles and the number of tillers that developed on the upper nodes. The number of internodes (NEN) was also counted. The weight of the main panicle (PPA) was measured using a digital scale.

The brix was measured at 3 stages on the third, fifth and seventh knots. A portable ATAGO® refractometer was used to measure brix.

Figure VI shows the portable refractometer, while Figure VII illustrates the steps involved in measuring the brix of the knots in each genotype.

Table I summarises the quantitative variables measured.

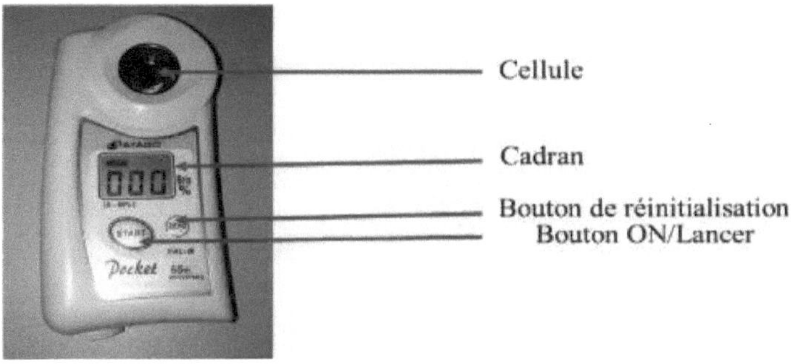

Figure VI: Illustration of the portable refractometer (BERE, 2019)

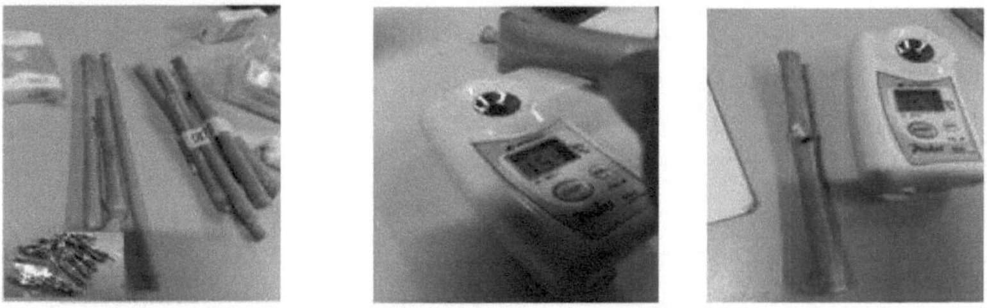

A-Selected nodes B-Deposit of a few drops C-Reading the results (15.4%)

Figure VII: Illustrations of the steps involved in measuring internode brix by genotype (BERE, 2019).

Figure VII: Illustrations of the steps involved in measuring the brix of knots by genotype (BERE, 2019).

Table I: List and abbreviations of quantitative variables measured

characters	abbreviations	Measurement unit	Stadium
Number of days to levee	NJL	day	levëe
Planting	TLV		Hard grain
Tallage^tile	TLU		Hard grain
Tallage aerien	TLA		Hard grain
Number of days swelling	NJG	day	Swelling
Number of sprouting days	NJE	day	Emergence
Number of days to flowering	NJF	day	Flowering
Length of the 3eme leaf under the panicle	LOF	cm	Pateux
Width of the 3eme leaf under the panicle	LAF	cm	Pateux
Number of entre-nreud	NEN		Pateux
Knot length	LOE	Cm	Pateux
Height of the plant	HPL	Cm	Pateux
Main stem diameter	DTP	Cm	Pateux
characters	**abbreviations**	**Measurement unit**	**Stadium**
Length of peduncle	LPE	Cm	Pateux
Panicle length	LOP	Cm	Pateux
Panicle width	LAP	Cm	Pateux
Main panicle weight	PPA	g	Pateux
Brix has 3 stages	BRIX	%	Milky, pasty, hard grains

III.3.2.2. Qualitative variables

The qualitative characteristics ëtudiës have ële numbered 07. The colour of the seedlings (CPL) was measured at the yeast stage and the flavour of the grains was measured at the pasty stage using an organoleptic test. The other qualitative variables, namely exsertion (EXE), dry grain shape (FGR), glume colour (CGL) and grain coverage (COG), were measured at the mature stage. Grain colour was measured at the hard grain stage.

Exsertion (EXE) is the distance between the top of the panicle leaf sheath and the first branches of the main panicle. The exsertion has ëlë considërëe positive or protruding when the panicle is complement dëgagëe of the sheath. In the case where the sheath embraces the panicle, the exsertion is negative. In the third case, there is no exsertion. Figure VIII below shows the sclu'ina of different sorghum exsertions.

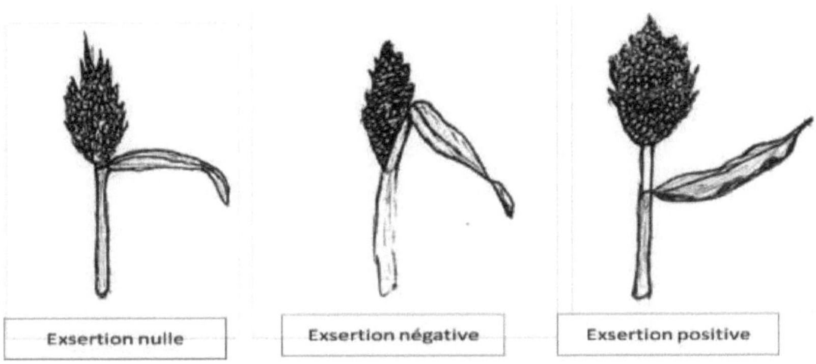

Figure VIII: Diagram of different sorghum exsertions (BERE, 2020)

Grain coverage (COG) is estimated on the basis of the proportion of grain wrapped in the glumes. The %, Ц, %, total and longer hull coverages have been defined. Figure IX below shows the sclK'ina of different sorghum grain covers.

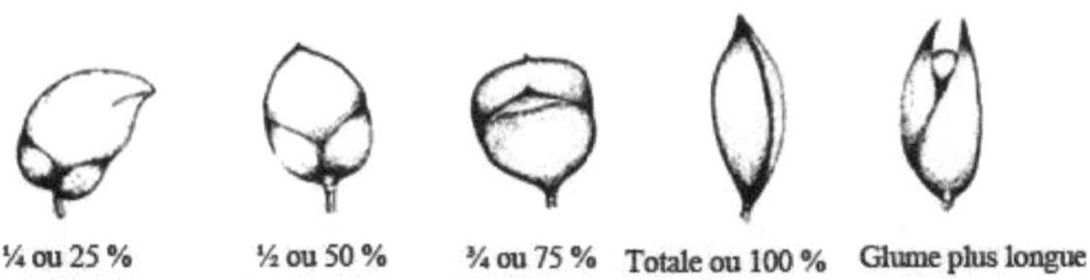

Figure IX: Diagram of different sorghum grain covers (IBPGR/ICRISAT, 1993)

The shape of the dry grain (FGR) has been defined according to two variants: the foveated shape (if the grain has a dimple) and the non-foveated shape (if the grain is convex). Figure X below shows a diagram of the different shapes of dry sorghum grain.

Figure X: Diagram of different shapes of dry sorghum grain (IBPGR/ICRISAT, 1993)

Table II summarises the qualitative variables measured

Table II: List and abbreviations of qualitative variables measured

characters	Abreviations	Stadium

Colour of seedlings	PLC	Levee
Exsertion	EXE	Maturity
Dry grain shape	FGR	Maturity
Colour of glumes	CGL	Maturity
Grain coverage	COG	Maturity
Grain flavour	SGR	Pateux
Grain colour	CGR	Maturity

III.4. 4. Data analysis methods

Descriptive statistics were calculated using Excel, XLSTAT and GENSTAT software. The means of the quantitative variables, the frequencies of the qualitative characteristics and the photopëriodism coefficient (Kp) were calculated using EXCEL 2016 software.

$$Kp = \frac{NJF_{D1} - NJF_{D2}}{\textit{Décalage de jours de la date 1 à la date 2}} ;$$ with $NJFDi$: Number of

Shift of days from date 1 to *date* 2

days to flowering of the first sowing date; $NJFD2$: Number of days to flowering of the second sowing date;

The Newman-Keuls test was ëtë rë^ë to determine the sëparation of the means of each trial using XLSTAT software version 2016.

One-factor and two-factor (interaction) analysis of variance (ANOVA) a ëtë rëalisëe using GENSTAT ëdition 4.lnk software.

CHAPTER III

RESULTS AND DISCUSSION

I. Results

I.1 Variation in qualitative characteristics

L1.1. Variation in quality characteristics at 1ere sowing date (29 June)

The results recorded in Table III present the distribution of gënotypes by variable according to modality and observation stage. In fact, at the emergence stage, 02 modalities were observed in terms of seedling colour, with a proportion of green-coloured seedlings of 55.172% and a proportion of purplish-coloured seedlings of 44.828%. Observations of glume colour revealed 06 distinct colours.

Table III: Frequencies of qualitative characteristics observed at different stages of the development cycle at the first sowing date (29 June)

Characters	Stadium	Details	Frequencies in
Exsertion (EXE)	Maturity	Positive	85,069%
		None	6,897%
		Negative	8,034%
Colour of seedlings (CPL)	Levee	purple	44,828%
		green	55,172%
Grain shape (FGR)	Maturity	Foveolee	73,552%
		Non foveolee	26,448%
Colour of glumes (CGL)	Maturity	light yellow	20,690%
		black	31,034%
		brown	18,379%
		dark brown	16,103%
		red	9,207%
		Spotted light-yellow	4,586%
Grain coverage (COG)	Maturity	G+L	3,448%
		100% covered	24,138%
		75% covered	44,828%
		50% covered	13,793%
		25% covered	13,793%
Grain colour (CGR)	Maturity	light red	4,586%
		dark red	44,828%
		mottled white	11,483%
		dirty white	39,069%
Grain flavour (SGR)	Pateux	No sugar (NS)	95,414%
		Legerement sucre (LS)	4,586%

***G+L**: hull longer than the grain.*

1.1.2. Variation in quality characteristics at 2eme sowing date (23 July)

The results presented in Table IV show the distribution of gënotypes by variable according to the modalitës and the observation stage. In fact, at the emergence stage, 02 modalitës were observed in terms of seedling colour, with a proportion of green-coloured seedlings of 55.172% and a proportion of purplish-coloured seedlings of 44.828%. Observations on the colour of the glumes at the ripening stage revealed 06 distinct colours.

Table IV: Frequencies of qualitative characteristics observed at different stages of the development cycle at the second sowing date (23 July)

Characters	Stadium	Details	Frequencies in
Exsertion (EXE)	Maturity	Positive	86,207%
		None	6,897%
		Negative	6,897%
Colour of seedlings (CPL)	Levee	purple	44,828%
		green	55,172%
Grain shape (FGR)	Maturity	Foveolee	72,414%
		Non foveolee	27,586%
Colour of glumes (CGL)	Maturity	light yellow	20,690%
		black	31,034%
		brown	17,241%
		dark brown	16,103%
		red black	9,207%
		Mottled yellow-light	5,725%
Grain coverage (COG)	Maturity	G+L	3,448%
		100% covered	24,138%
		75% covered	44,828%
		50% covered	13,793%
		25% covered	13,793%
Grain colour (CGR)	Maturity	light red	3,448%
		dark red	44,828%
		mottled white	13,793%
		dirty white	37,931%
Grain flavour (SGR)	Pateux	No sugar (NS)	96,552%
		Legerement sucre (LS)	3,448%

***JCT**: light yellow spotted;* ***G+L**: husk longer than the grain.*

Figure XI illustrates the shape of the sweet stalk sorghum grain and Figure XII shows the different colours of the sweet stalk sorghum grains we studied.

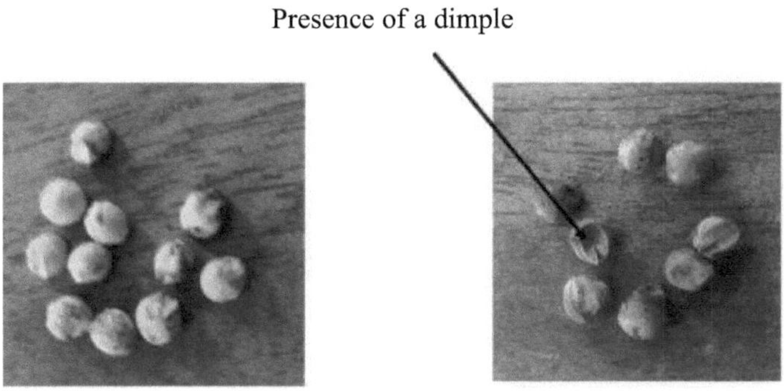

A-grains not foveolated A-foveolate seeds

Figure XI: Illustration of the grain shape of sweet-stem sorghum (BERE, 2019)

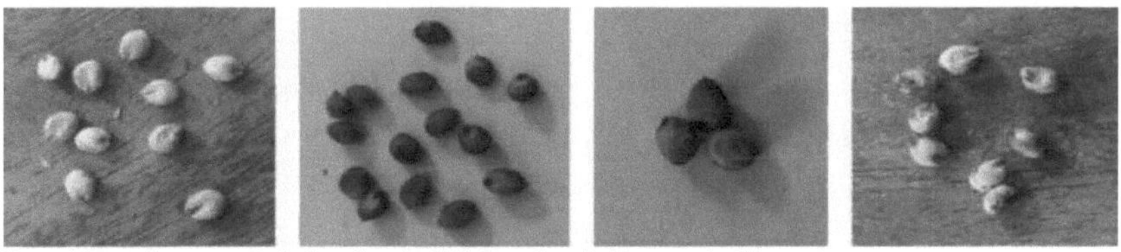

A-dirty white grain B-dark red grain C-light red grain D-spotted white grain

Figure XII: Illustration of the colour of sweet stem sorghum grains (BERE, 2019)

1.1.3. Effect of sowing date on variation in quality characteristics

Table V shows the minimum, maximum and mean numbers; the F probabilities; the coefficients of variation; and the standard errors by mode for the categorical variables.

The results of the analysis of variance (ANOVA) carried out on the numbers of the different modalities per qualitative variable show that the sowing date had no significant influence on the expression of the qualitative parameters. The coefficients of variation (CV) were low in most cases. This would indicate that the sowing date had very little influence on these characteristics.

Table V: Results of the analysis of variance (ANOVA) on the numbers of qualitative variables

		WORKFORCE								
Minimum				*Maxima*		*Averages*				
Characters	*Details*	*D1*	*D2*	*D1*	*D2*	*D1*	*D2*	*F pr.*	*CV*	*Visit*
	Positive	24	25	25	25	24,67	25	**0.374 ns**	1,64	0,408
EXE	None	2	2	2	2	2	2	-	0	0,000
	Negative	2	2	3	2	2,33	2	**0.374 ns**	18,84	0,408

PLC	purple	14	14	16	16	14,66	14,66	**1 ns**	7,87	1,155
	green	13	13	15	15	14,33	14,33	**1 ns**	8,06	1,155
FGR	Foveole	21	21	22	21	21,33	21	**0.374 ns**	1,93	0,408
	No foveole	7	8	8	8	7,67	8	**0.374 ns**	5,21	0,408
CGL	light yellow	6	6	6	6	6	6	-	0	0,000
	black	9	9	9	9	9	9	-	0	0,000
	brown	5	5	6	5	5,33	5	**0.374 ns**	7,9	0,408
	MF	4	4	6	6	4,67	4,67	**1 ns**	24,74	1,155
	red	2	2	3	3	2,67	2,67	**1 ns**	21,65	0,577
	JCT	1	1	2	2	1,33	1,67	**0.519 ns**	38,49	0,577
COG	G+L	1	1	1	1	1	1	-	0	0,000
	100%	7	7	7	7	7	7	-	0	0,000
	75%	13	13	13	13	13	13	-	0	0,000
	50%	4	4	4	4	4	4	-	0	0,000
	25%	4	4	4	4	4	4	-	0	0,000
CGR	light red	1	1	2	1	1,33	1	**0.374 ns**	34,99	0,408
	dark red	13	13	13	13	13	13	-	0	0,000
	BT	2	4	4	4	3,33	4	**0.374 ns**	22,27	0,817
	dirty white	11	11	12	11	11,33	11	**0.374 ns**	3,66	0,408
SGR	LS	1	1	2	1	1,33	1	**0.374 ns**	34,99	0,408
	NS	27	28	28	28	27,67	28	**0.374 ns**	1,47	0,408

***EXE** : exsertion ; **CPL** : seedling colour ; **FGR** : grain shape ; **CGL** : glume colour ; **COG** : grain coverage ; **CGR** : grain colour ; **SGR** : grain flavour ; **D1** : first sowing date (29/06/2019) ; **D2** : second sowing date (23/07/2019) ; **CV :** coefficient of variation ; **Se** : standard error ; **F pr.** : F probability ; **ns** : not significant ; **-** : no value ; **JCT** : light yellow spotted ; **G+L** : husk longer than the grain ; **LS** : slightly sweet ; **NS** : not sweet ; **MF** : dark brown ; **BT** : white spotted.*

I.2 Variation in quantitative characteristics

1.2.1. Average performance by variable and sowing date of the genotypes studied

At the first sowing date (Table VI), the results of the analysis of variance indicate that only the variables number of days in the ground (NJL) and main stem diameter (DTP) showed non-significant differences between genotypes. However, the 18 other variables showed significant differences between the genotypes evaluated. The number of days from sowing to flowering (NJF) ranged from 76.33 days for the earliest genotype (genotype 124) to 96.67 days for the latest genotype (genotype 89). Brix a varied from 13.42% (genotype 02) to

19.13% (genotype 125) with an average of 16.13%. Plant height ranged from 193.10 cm (genotype 104) to 417.20 cm (genotype 120) with an average of 334.94 cm.

At the second sowing date (Table VII), 19 variables showed significant differences between genotypes while only one variable, main stem diameter (MSD), showed no significant differences between genotypes. Days to flowering (DFT) ranged from 67.33 days for the earliest genotype (genotype 97) to 81.33 days for the latest genotype (genotype 79). Brix varied from 13.94% (genotype 108) to 20.51% (genotype 70) with an average of 16.87%. Plant height ranged from 142.90 cm (genotype 104) to 322.4 cm (genotype 115) with an average of 249.25 cm.

Table VI: Variability and average performance of each genotype by variable at the first sowing date

Variables

Genotypes	*TLU*	*TLV*	*TLA*	*PPA*	*NJL*	*NJG*	*NJF*	*NJE*	*NEN*	*LPE*	*LOP*	*LOF*	*LOE*	*LAP*	*LAF*	*HPL*	*DTP*	*BSP*	*BSL*	*BGD*
2	0,00	0,00	5,00	160,40	4,00	72,50	81,50	79,00	13,75	60,58	35,83	64,92	40,58	11,25	7,86	330,00	21,17	11,35	14,30	14,62
70	1,67	2,00	7,67	125,20	3,00	77,67	85,67	80,00	12,56	52,08	23,04	66,62	46,78	6,17	8,30	303,30	19,67	18,31	16,32	22,08
73	0,00	0,00	1,67	151,90	4,00	83,67	93,67	89,00	14,83	47,80	32,06	68,00	46,59	8,67	9,66	345,20	21,50	16,53	16,50	18,84
79	1,00	1,00	4,67	129,00	3,33	80,33	87,33	83,33	14,44	33,61	13,81	57,72	45,94	7,39	8,34	290,60	21,78	14,87	16,34	15,82
81	3,00	3,67	5,33	148,50	3,33	89,00	95,67	93,67	12,56	46,17	39,03	74,72	56,92	9,39	6,92	362,30	23,11	14,41	16,73	16,41
85	2,67	3,00	3,67	112,60	3,00	89,00	94,67	91,67	13,11	46,80	36,00	73,00	46,33	9,39	6,06	350,30	23,06	14,11	16,57	15,17
88	0,67	0,67	4,67	134,50	3,33	78,33	86,00	84,00	14,11	33,39	13,89	60,33	45,82	7,33	8,47	288,60	20,00	16,60	13,82	17,04
89	2,00	2,67	5,33	160,00	4,00	85,33	96,67	93,33	13,61	42,61	38,78	76,28	51,14	10,78	6,69	346,40	27,28	16,49	14,48	17,30
92	4,00	4,33	11,00	154,80	3,00	78,00	85,00	81,00	11,78	75,89	37,50	80,22	51,44	8,67	7,63	382,90	21,78	12,78	14,87	19,09
97	2,00	2,00	8,00	143,90	3,00	74,67	83,00	77,67	12,11	55,79	28,89	75,50	48,99	7,41	8,33	327,30	19,28	16,59	19,09	18,10
98	1,00	1,00	8,67	136,80	3,33	72,33	79,33	76,00	12,08	50,04	30,00	74,22	46,60	7,56	7,99	328,60	21,00	15,18	14,82	21,44
103	1,67	2,00	2,00	163,00	4,00	85,33	94,00	90,00	15,89	34,78	30,11	75,22	36,33	9,33	10,26	303,60	26,11	15,99	15,47	19,00
104	0,33	0,33	7,00	171,00	3,33	80,00	87,00	83,33	11,67	36,67	24,89	69,06	23,50	9,11	9,58	193,10	18,67	12,41	19,10	16,66
105	0,67	1,67	22,00	159,40	3,00	69,33	78,67	73,67	11,67	66,06	36,24	72,56	49,67	8,28	7,86	359,90	21,33	14,52	20,54	18,41
108	2,67	3,33	7,00	142,70	3,33	73,00	79,33	75,00	11,22	55,61	32,56	73,56	52,04	8,89	6,00	320,60	21,39	13,38	12,47	16,20
109	2,33	3,33	2,00	143,50	3,33	80,00	86,00	82,00	13,44	52,56	30,44	67,28	49,78	9,72	8,23	353,80	21,11	14,92	15,00	18,04
110	2,33	3,33	4,00	148,40	3,00	83,67	92,67	85,67	13,67	48,56	41,67	81,50	49,53	9,72	8,04	351,30	20,17	14,83	16,31	18,07
112	3,33	3,67	11,00	148,90	3,33	79,00	90,00	85,00	13,56	71,17	29,98	61,94	49,78	7,28	6,74	408,30	18,78	16,09	15,26	18,38
115	0,33	0,33	3,33	155,00	3,33	84,00	94,67	90,33	14,00	74,67	33,61	66,72	54,33	10,00	6,39	415,00	20,56	12,50	14,04	20,52

119	2,33	2,33	3,67	166,00	3,67	84,33	91,00	87,00	15,33	51,94	41,11	75,33	43,21	10,78	8,34	390,30	22,78	14,75	15,53	16,67
120	3,33	4,67	6,33	146,40	3,33	76,33	86,67	81,67	13,67	77,33	38,78	61,67	56,04	8,39	6,29	417,20	20,11	14,27	17,40	15,27
123	1,33	2,33	1,00	133,20	3,67	79,67	87,00	83,33	13,33	50,22	30,50	68,33	51,89	8,94	8,20	350,60	18,67	14,99	15,43	17,97
124	2,33	2,67	8,67	147,80	3,00	71,33	76,33	73,67	13,56	51,33	25,61	66,28	43,89	8,83	8,41	320,60	20,83	14,02	18,13	18,82
125	1,33	1,33	17,00	155,40	3,00	73,67	79,00	76,00	12,22	68,83	36,39	68,92	49,33	7,94	7,18	371,10	21,56	18,30	18,83	20,27
127	2,33	3,00	8,67	170,00	3,67	73,33	78,67	75,67	12,56	46,38	28,44	69,14	26,00	13,22	9,43	216,90	19,44	8,29	14,76	19,51
129	2,67	3,00	3,33	146,60	3,67	80,67	92,33	85,00	15,17	50,56	34,22	68,28	43,56	9,50	8,43	345,80	19,72	15,32	16,02	20,30
130	2,67	3,67	2,33	130,10	3,67	81,33	90,67	85,67	14,78	49,06	31,81	69,78	42,61	9,17	8,76	351,70	21,56	15,90	13,91	16,44
136	2,00	3,00	9,00	143,60	3,33	71,33	79,00	74,33	9,44	52,17	33,94	75,33	56,72	9,22	5,76	296,40	20,22	11,94	12,64	18,34
137	3,67	4,67	10,00	129,10	3,33	66,33	78,33	73,67	9,33	54,67	33,33	74,11	56,41	9,00	5,42	283,90	19,89	13,59	13,87	16,03
F pr	0,015	0,022	0,0003	0,028	0,110	$<10^{-4}$	$<10^{-4}$	$<10^{-4}$	$<10^{-4}$	$<10^{-4}$	$<10^{-4}$	$<10^{-4}$	$<10^{-4}$	0,002	$<10^{-4}$	$<10^{-4}$	0,069	$<10^{-4}$	$<10^{-4}$	$<10^{-4}$
significance	*	*	***	*	ns	***	***	***	***	***	***	***	***	**	***	***	ns	***	***	***

*TLV: plant tillering height; TLU: useful tillering height; TLA: aerial tillering height; NJL: number of days to 50% shoot emergence; NJG: number of days to 50% plant swelling per line; NJE: number of days to 50%> plant heading per line; NJF: number of days to 50%> flowering of plants per line; LOF: length of the 3eme sub-panicular leaf; LAF: width of the 3eme sub-panicular leaf; NEN: number of internodes; LOE: length of internodes; HPL: height of the plant; DTP: diameter of the main stem; LPE: length of the peduncle; LOP: length of the main panicle; LAP: width of the main panicle; PPA : main panicle weight; BSL: brix at milky stage; BSP: brix at heavy stage; BGD: brix at hard grain stage; ns: not significant; *: significant; **: highly* significant24 ****; tris highly significant.*

Table VII: Variability and average performance of each genotype by variable at the second sowing date

Variables

Genotypes	*TLU*	*TLV*	*TLA*	*PPA*	*NJL*	*NJG*	*NJF*	*NJE*	*NEN*	*LPE*	*LOP*	*LOF*	*LOE*	*LAP*	*LAF*	*HPL*	*DTP*	*BSP*	*BSL*	*BGD*
2	2,00	2,67	4,67	152,20	3,67	65,67	75,00	71,33	9,89	58,88	31,46	65,11	44,46	8,88	7,41	254,00	17,00	15,32	12,83	18,84
70	3,00	3,67	6,00	130,20	3,67	65,67	72,33	68,67	8,33	49,00	21,10	71,22	48,94	5,89	7,99	229,90	15,89	21,24	17,94	22,36
73	3,33	4,33	2,67	137,50	4,00	72,67	80,00	76,67	11,67	51,68	27,89	63,22	41,42	7,29	13,24	259,30	16,83	18,22	17,17	18,97
79	2,33	2,67	2,00	128,70	3,00	73,33	81,33	79,33	11,67	37,19	15,15	57,56	39,63	7,34	6,34	230,40	15,89	15,81	13,77	18,36
81	3,67	5,67	2,33	146,5	3,67	69,6	80,3	77,0	9,22	46,6	27,4	65,56	51,4	6,47	4,20	251,2	16,2	15,0	14,7	18,0

				0		7	3	0		7	3		4			0	8	4	0	4
85	4,00	8,33	2,00	148,60	3,00	65,67	74,00	69,67	8,67	46,28	28,72	64,33	47,29	7,58	4,41	231,90	15,39	15,43	12,31	17,28
88	3,00	3,00	2,00	118,90	3,67	67,00	76,33	74,00	12,33	30,74	14,12	58,56	46,06	5,99	5,64	246,80	19,00	17,14	15,09	17,14
89	3,67	6,33	6,33	142,00	4,00	67,33	79,33	74,67	9,11	46,44	30,72	64,89	48,23	7,40	4,92	239,00	18,28	14,03	14,02	20,02
92	3,00	4,33	6,33	170,80	3,33	65,00	73,67	68,67	8,22	72,22	33,29	68,78	54,59	10,36	5,24	277,10	16,17	17,00	14,92	18,53
97	3,67	6,00	7,67	149,10	3,00	59,33	67,33	63,67	7,56	50,77	31,78	68,78	51,17	8,49	6,49	225,10	16,00	19,29	16,30	22,96
98	3,67	5,33	4,33	154,10	3,67	65,33	71,33	68,33	7,33	53,13	29,77	67,11	50,33	7,65	5,78	222,40	14,00	20,10	14,74	21,69
103	3,67	5,00	3,00	131,70	4,00	70,00	79,67	74,67	10,89	43,12	29,99	64,78	35,73	7,07	7,20	225,80	18,44	16,36	14,13	16,59
104	0,00	0,00	3,00	126,30	3,00	69,33	79,33	75,00	10,78	30,52	28,66	71,11	19,56	7,31	8,98	142,90	30,78	15,33	14,46	15,29
105	3,00	4,67	7,67	141,40	3,00	59,67	67,67	63,33	7,56	67,02	31,11	64,56	57,46	6,82	6,17	269,30	14,06	18,64	16,80	19,89
108	3,67	7,33	7,67	131,40	3,00	66,00	74,67	69,67	7,22	49,62	30,17	61,44	51,66	8,26	4,30	210,20	15,00	13,49	12,90	15,44
109	3,00	5,00	4,00	120,30	3,67	68,67	79,33	73,33	9,33	46,33	24,67	62,89	47,64	10,64	6,43	244,40	17,17	16,06	14,14	19,36
110	4,00	6,00	4,33	148,80	3,67	65,67	77,67	71,00	10,44	52,49	35,86	74,17	45,71	7,90	6,49	269,70	16,94	17,92	15,98	17,32
112	4,00	4,67	6,33	129,70	3,33	65,33	74,00	70,00	9,67	66,99	27,11	67,22	52,07	10,85	5,39	294,90	15,72	20,31	15,17	21,84
115	3,67	4,67	4,00	145,50	3,33	69,67	79,67	75,67	10,56	69,64	28,83	70,67	51,71	8,29	5,29	322,30	16,33	17,25	15,69	20,19
119	2,00	4,00	4,67	144,90	3,67	68,33	78,00	75,00	10,56	53,39	32,42	71,67	46,20	10,80	6,41	276,40	17,67	14,58	15,06	19,37
120	4,67	6,00	4,67	146,60	3,33	67,67	79,33	74,00	10,00	71,11	40,69	63,44	54,76	10,44	5,77	310,60	17,56	12,69	11,49	20,68
123	1,67	4,67	2,33	131,70	4,00	67,33	77,00	73,00	10,89	55,58	29,11	66,67	49,07	7,28	7,10	288,40	16,78	15,62	15,94	16,83
124	4,67	6,67	7,00	119,90	3,00	66,67	73,33	69,67	9,78	48,19	23,55	67,89	43,51	6,89	6,92	238,00	15,33	17,82	14,09	18,97
125	3,33	5,33	11,33	157,80	4,00	62,67	68,00	65,00	8,33	66,97	40,00	64,78	56,88	9,75	6,16	284,30	14,78	20,24	15,40	21,68
127	2,00	2,33	5,33	150,60	4,00	63,00	69,67	66,00	10,11	39,83	32,89	77,94	23,21	7,88	8,21	177,80	18,28	18,26	14,67	20,44
129	3,00	4,33	3,00	142,00	3,33	67,67	78,67	75,00	12,44	55,32	30,34	59,44	39,60	7,54	7,08	285,60	19,61	16,99	15,18	20,34

130	2,67	4,67	2,33	130,30	3,67	68,67	81,00	74,67	11,56	57,11	33,78	62,33	43,79	8,18	6,83	277,60	18,39	15,38	16,09	21,88
136	5,00	6,33	9,67	144,50	3,67	63,33	71,67	66,33	7,67	55,24	28,45	61,56	52,64	7,21	4,69	222,10	14,00	15.84	11,21	17,10
137	4,33	6,00	10,67	156,30	3,67	61,67	67,67	64,67	7,44	51,87	27,46	63,89	52,74	6,14	4,32	220,70	15,33	15,90	13,43	17,12
Fpr	0,004	0,024	0,043	0,011	0,007	<10'4	<10'4	<10'4	<10'4	<10'4	<10'4	0,0001	<10'4	<10'4	0,038	<10'4	0,587	<10'4	<10'4	<10'4
significance	**	**	**	*	**	***	***	***	***	***	***	***	***	***	*	***	**ns**	***	***	***

*TLV: vegetative tillering; TLU: useful tillering; TLA: aerial tillering; NJL: number of days to 50% bud break; NJG: number of days to 50% swelling per line; NJE: number of days to 50%> heading per line; NJF: number of days to 50%> flowering of plants per line; LOF: length of the 3eme sub-panicular leaf; LAF: width of the 3eme sub-panicular leaf; NEN: number of internodes; LOE: length of internodes; HPL: height of the plant; DTP: diameter of the main stem; LPE: length of the peduncle; LOP: length of the main panicle; LAP: width of the main panicle; PPA : main panicle weight; BSL: brix at milky stage; BSP: brix at pasty stage; BGD: brix at hard grain stage; ns: not significant; *: significant; **: highly significant; ***: very highly significant.*

1.2.2. Effect of sowing date on the expression of quantitative variables

The results of the analysis of variance presented in Table VIII, taking into account the sowing date, showed non-significant differences between the gënotypes for the variables peduncle length (LPE) and internode length (LOE). However, the differences were significant for the variables main panicle weight (MPW), number of days to emergence (NJL), highly significant for aerial tillering (TLA) and very highly significant for the rest of the variables including height, brix and cycle.

For the sowing date x genotype interaction, the effect was non-significant on genotypes for vegetative, useful and aerial tillering (TLV, TLU, TLA), number of days to emergence (NJL), leaf width (LAF), plant height (HPL) and main stem diameter (DTP). On the expression of the other variables, the effect was significant for main panicle weight (PPA), number of internodes (NEN), peduncle length (LPE) and internode length (LOE); highly significant on brix at hard grain stage (BGD) and very highly significant on sowing swelling (NJG), sowing flowering (NJF), sowing heading (NJE), panicle length (LOP) and leaf length (LOF), panicle width (LAP) and brix at milky (BSL) and pasty (BSP) stages.

Table VIII: Results of analysis of variance (ANOVA) with one factor (sowing date and genotype) and two factors (sowing date x genotype)

variables	*F pr. date*	*Fpr. genotype*	*Fpr. Date x genotype*	*CV as % of total*	*go to*
TLV	<0,001 ***	<0,001 ***	0.370 **ns**	49,880	1,800
TLU	<0,001 ***	<0,001 ***	0.381 **ns**	52,140	1,345
TLA	0,009 **	<0,001 ***	0.575 **ns**	70,060	4,139
PPA	0,016 *	0,002 **	0,041 *	11,650	16,740

NJL	0,042 *	<0,001 ***	0.794 **ns**	12,820	0,442
NJG	<0,001 ***	<0,001 ***	<0,001 ***	2,260	1,635
NJF	<0,001 ***	<0,001 ***	<0,001 ***	2,010	1,628
NJE	<0,001 ***	<0,001 ***	<0,001 ***	2,280	1,750
NEN	<0,001 ***	<0,001 ***	0,042 *	8,730	0,9910
LPE	0.519 **ns**	<0,001 ***	0,019 *	7,91	4,169
Variables	**F pr. date**	**F pr. genotype**	**F pr. Date x genotype**	**CV as % of total**	**go to**
LOP	<0,001 ***	<0,001 ***	<0,001 ***	9,59	2,927
LOF	<0,001 ***	<0,001 ***	<0,001 ***	6,38	4,346
LOE	0.367 **ns**	<0,001 ***	0,042 *	7,670	3,584
LAP	<0,001 ***	<0,001 ***	<0,001 ***	15,51	1,326
LAF	<0,001 ***	<0.001 ***	0.869 **ns**	24,28	1,721
HPL	<0,001 ***	<0,001 ***	0.028 **ns**	7,40	21,62
DTP	<0,001 ***	0.569 **ns**	0.279 **ns**	22,74	4,334
BSP	<0,001 ***	<0,001 ***	<0,001 ***	10,98	1,725
BSL	<0,001 ***	<0,001 ***	<0,001 ***	9,18	1,400
BGD	<0,001 ***	<0,001 ***	0,001 **	9,15	1,697

ns: *not significant;* ****:*** *significant;* *****:*** *highly significant;* ******:*** *very highly significant;* ***Cv:*** *coefficient of variation;* ***se:*** *standard error.*

1.2.3. Comparison of genotype performance by variable as a function of sowing date

Table IX shows the performance of the gënotypes by sowing date. Variables such as tallage vëgëtatif (TLV) and utile (TLU), number of days to levëe (NJL), brix at bare (BSP) and hard grain (BGD) stages showed higher values at the second sowing date compared with the first sowing date. Late sowing would result in an increase in the number of days to emergence (+0.14 days) and a higher production of v®g®tative tillers (+2.44 tillers), useful tillers (+1.3 tillers), brix at the pale stage (+2) and hard grain (+1.16) than early sowing. However, advanced sowing would result in a reduction in tillering (-1.31), number of days to swelling (-11.89), number of days to ®piaison (-11.05), number of days to flowering (-11.06), leaf length (-4.35), leaf width (-1,4), number of internodes (-3.46), plant height (-85.69), main stem diameter (4.13), panicle length (-2.63), panicle width (-1.05), panicle weight (-6.19) and brix at milky stage (-1.14). There were no significant differences between the two sowing dates for internode length and panicle length.

Table IX: Average performance of genotypes by sowing date

VARIABLES -	*MINIMA*	*MAXIMA*	*AVERAGE*	*F PR*	*DIFFERENCES*

	D1	D2	D1	D2	D1	D2	*(SOWING DATE FACTOR)*	*IN AVERAGES (D2-D1)*
TLV	0,00	0,00	8,00	11,00	2,39	4,83	<0.001 ***	2,44
TLU	0,00	0,00	5,00	6,00	1,93	3,23	<0.001 ***	1,3
TLA	0,00	0,00	22,00	17,00	6,39	5,08	0.009 **	-1,31
NJL	3,00	3,00	4,00	4,00	3,38	3,52	0.042 *	0,14
NJG	64,00	58,00	90,00	74,00	78,37	66,48	<0.001 ***	-11,89
NJE	71,00	62,00	94,00	81,00	82,36	71,31	<0.001 ***	-11,05
NJF	75,00	66,00	97,00	82,00	86,49	75,43	<0.001 ***	-11,06
LOF	55,67	52,67	88,50	80,83	70,27	65,92	<0.001 ***	-4,35
LAF	5,17	3,30	10,73	27,07	7,79	6,39	<0.001 ***	-1,4
NEN	9,00	6,67	16,67	13,67	13,09	9,63	<0.001 ***	-3,46
LOE	10,34	9,67	31,50	31,07	23,48	23,24	0.367 **ns**	-0,24
HPL	188,33	129,00	464,33	347,67	334,94	249,25	<0.001 ***	-85,69
DTP	14,33	10,50	29,67	61,67	21,13	17,00	<0.001 ***	-4,13
LPE	29,67	25,83	80,83	79,50	52,94	52,53	0.519 **ns**	-0,41
LOP	13,27	11,67	45,67	41,67	31,82	29,19	<0.001 ***	-2,63
LAP	5,33	5,24	17,00	11,67	9,07	8,02	<0.001 ***	-1,05
PPA	107,00	101,21	205,97	189,16	146,82	140,63	0.016 *	-6,19
BSL	10,83	9,60	20,70	18,20	15,82	14,68	<0.001 ***	-1,14
BSP	6,57	8,83	19,63	22,63	14,60	16,80	<0.001 ***	2,2
BGD	13,20	12,80	23,83	24,17	17,96	19,12	<0.001 ***	1,16

***D1** :first sowing date (29/06/2019) ; **D2** : second sowing date (23/07/2019) ; **ns** : not significant; * : significant; ** : highly significant; *** : very highly significant; **Cv** : coefficient of variation.*

1.2.4. Cycle reduction and photoperiodism coefficient

Table X shows that all the accessions responded with a reduction in the cycle ranging from 03 to about 21 days for a 24-day difference between the two sowing dates. All gënotypes were sensitive to variation in photopëriod. Five (05) gënotypes were ë1.ë weakly sensitive (Kp < 0.3), 12 gënotypes have ëlë moderately sensitive (0.3 < Kp < 0.5), 11 have ëlë highly sensitive (0.5 < Kp < 0.8) and only one gënotype has ëlë very highly sensitive (Kp > 0.8). The 124 gënotype has the lowest coefficient of photopëriodism (Kp = 0.125) while the 85 gënotype has the highest coefficient (Kp = 0.861).

Table X: Coefficient of photoperiodism and genotype sensitivity

GENOTYPES	*NJFD1*	*NJFD2*	*RDC*	*KP*	*SENSITIVITY*
124	76,33	73,33	3,00	0,125	FS
108	79,33	74,67	4,67	0,194	FS

2	80,00	75,00	5,00	0,208	FS
79	87,33	81,33	6,00	0,250	FS
109	86,00	79,33	6,67	0,278	FS
136	79,00	71,67	7,33	0,306	MS
120	86,67	79,33	7,33	0,306	MS
104	87,00	79,33	7,67	0,319	MS
98	79,33	71,33	8,00	0,333	MS
127	78,67	69,67	9,00	0,375	MS
88	86,00	76,33	9,67	0,403	MS
130	90,67	81,00	9,67	0,403	MS
123	87,00	77,00	10,00	0,417	MS
137	78,33	67,67	10,67	0,444	MS
105	78,67	67,67	11,00	0,458	MS
125	79,00	68,00	11,00	0,458	MS
92	85,00	73,67	11,33	0,472	MS
119	91,00	78,00	13,00	0,542	HS
70	85,67	72,33	13,33	0,556	HS
127	92,33	78,67	13,67	0,569	HS
73	93,67	80,00	13,67	0,569	HS
103	94,00	79,67	14,33	0,597	HS
110	92,67	77,67	15,00	0,625	HS
115	94,67	79,67	15,00	0,625	HS
81	95,67	80,33	15,33	0,639	HS
97	83,00	67,33	15,67	0,653	HS
112	90,00	74,00	16,00	0,667	HS
89	96,67	79,33	17,33	0,722	HS
85	94,67	74,00	20,67	0,861	HRT

NJFD1: average number of days 50% flowering of the first sowing date (29/06/2019) ; NJFD2: average number of days 50% flowering of the second sowing date (23/07/2019) ; ***RDC:*** *number of days of cycle reduction ;* ***Kp*** *: coefficient of photoperiodism;* ***FS:*** *weakly susceptible (for* ***Kp < 0.3****);* ***MS:*** *moderately susceptible (for* ***0.3 < Kp < 0.5****);* ***HS:*** *highly susceptible (for* ***0.5 < Kp < 0.8****);* ***THS:*** *very highly susceptible (for* ***0.8 < Kp < 1****).*

II. Discussion

The 24-day sowing interval produced no significant effect on the qualitative variables. The low coefficients of variation notedës indicate a modërëe variation in the qualitative characteristics between the two sowing dates. These results could be explained by the fact that the qualitative variables are generally governed by a very small number of genes that are generally little influenced by the environment. Qualitative characteristics are monogenic.

The sowing offset also did not significantly affect the length of the internode (LOE) and the length of the peduncle (LPE). However, it did cause a significant increase in the performance of the accessions for variables such as number of days emerged (NJL), vegetative and useful tillering (TLV and TLU), brix at the dull stage (BSP) and at the hard grain stage (BGD) and a significant drop in performance for the rest of the quantitative variables. In fact, plant height (HPL) decreased very significantly with the delay in sowing. Similar results are reported by NAMOANO (2017) who worked on 12 sweet grain sorghum genotypes. The reduction in size at the second sowing date is not explained by a reduction in internode length but rather by a reduction in the number of internodes. According to BEZOT (1963) and CLERGET *et al* (2008b), the later the sowing, the lower the plant height. The internodes are generated by the phytomeres, which are the growth units. This reduction in height results directly from the reduction in the number of phytomeres (internodes) produced, but rarely from their size.

The results also indicate that late sowings take slightly longer to emerge compared to early sowings. GNANSSOUNOU *et al* (2004) reported that the coleoptile of sugar sorghum appears above the soil after three (03) to four (04) days when the soil temperature is around 30°C. The increase in the number of days to emergence of delayed seedlings would be due to the temperature of the environment which is lower in July (28°C) than in June (30.9°C). According to OUEDRAOGO (2014), when a seed is buried in moist soil, it soaks up water and swells. Germination occurs rapidly in warm soils, with the coleoptile appearing first above the soil after three to four days, and takes longer, up to 10 days in colder soils (13 to 20°C).

The 24-day delay in sowing resulted in a reduction in the sowing-flowering cycle from 03 to approximately 21 days at the second sowing date. This result can be explained by the fact that all the genotypes studied are photoperiodic, as the reduction in the sowing-flowering cycle at the second sowing date is characteristic of photoperiodic plants. This immediately resulted in group flowering towards the end of the rainy season. GARNER and ALLARD (1923a), COCHEME and FRANQUIN (1967), CURTIS (1968) have indeed shown that the photopëriod sensitivity (photopëriodism) of sorghums causes a shortening of the cycle when sowing is delayed, thus favouring group flowering with the end of the rainy season. Our results are similar to those of GAPILI (2016) on ordinary sorghum and NAMOANO (2017) on sweet grain sorghum, who all found a reduction in the cycle ranging from 04 hours to 24 days and 03 to 10 days respectively. The time lag was 14 days for NAMOANO (2017) and 03 weeks for GAPILI *et al.* (2015). Thus, as argued by TRAORE *et al.* (2000), the reduction of the cycle is a characteristic of photoperiodic plants, which interrupt vegetative development in favour of seed production. This interruption is made without taking into account the complete development of the plant, but acts as a distress signal that intervenes to allow the transition from the vegetative phase to the reproductive phase in order to ensure the survival of the species.

The staggering of sowing dates shows an increase in the number of vegetative tillers (TLV) and useful tillers (TLU) and a decrease in aerial tillers (TLA) when sowing is delayed. The rapid interruption of the vegetative phase by the reproductive phase during delayed sowing would lead to a sort of vegetative compensation by the plant producing numerous vegetative tillers. According to DOGGETT (1988), the ability to till depends as much on the variety as on environmental conditions, in this case population density, nitrogen supply, temperature and photoperiod.

Data on brix showed that at the milky stage, late sowings produced less brix than early sowings, while at the pale and hard grain stages, late sowings produced higher brix than early sowings. The high brix at the milky stage (BSL) in early sowings is explained by the good phenology of the plants, which in turn is dictated by the sowing-flowering cycle, plant height and probably photoperiod. According to GUTJHAR (2012), high sugar and dry matter contents are due to good plant phenology. They are strongly favoured by a large number of large knots, the length of the sowing-flowering cycle and photoperiodism. The brix values were high at the pasty stage and at the hard grain stage for late sowings. These results show that the good phenology of the plants is not sufficient to explain the high sugar content of the first sowing date, but the rainfall must also be taken into account, as the brix measurements at the chalky and hard grain stages of the second sowing date took place during a dry period. The drought would have caused a significant drop in the amount of water in the plant and therefore, logically, an increase in the sugar content of the genotypes (SAWADOGO, 2013).

This study showed that all 29 tested genotypes of Burkina sweet stem sorghum are sensitive to photoperiod with a coefficient of photoperiodism varying from one genotype to another and ranging from 0.125 to 0.861. The genotypes shorten their sowing-flowering cycle by between 3 and 21 days for a sowing offset of 24 days. Genotype 124 was the least photoperiodic (Kp = 0.125) while genotype 85 was the most photoperiodic (Kp = 0.861).

In addition to phenological characteristics, delayed sowing significantly affects the expression of most vegetative characteristics and brix. However, it does not have a significant effect on quality characteristics.

The sowing date x genotype interaction also had a significant effect on most quantitative characteristics. In fact, the variables NJG, NJF, NJE, LOP, LOF and LAP showed very highly significant differences with values less than 0.001.

The genotypes could be used in sorghum breeding programmes. Biochemical and pathological characterisation of the genotypes identified could determine the effect of staggered sowing dates on juice composition and identify harmful micro-organisms.

Participatory and multifocal evaluation trials in several contrasting climatic areas, as well as off-season evaluation, could also help identify the genotypes of interest to producers and consumers and contribute to assessing the brix performance of sweet-stem sorghum and the stability of the genotypes.

BIBLIOGRAPHY

ECOFIN AGENCY, 2018. Burkina Faso: production cërëaHere reached 4 million tonnes in 2017/2018.

AHMADI N., CHANTEREAU J., LETHEQUE H. C., MARCHAND J. L., OUENDEBA B., 2002. Les cërëales. *In: Memento de I'agronome*. CIRAD and GRET, ISBN 2-86844-1297, 811822.

ANDREWS D. J., 1973. Effects of date of sowing on photosensitive Nigerian sorghums. *Experimental Agriculture,* **9**, 337-346.

ASIEDU J. J., 1989. *Transformation des produits agricoles en zone tropicale: approche technologique*. Edition Karthala et CTA, ISBN 2-86537-3347, 335p.

BARRO-KONDOMBO C. P., 2004. *Evaluation de la diversite genetique des sorghos des régions agricoles du centre-ouest et de la Boucle du Mouhoun*. Mëmoire de DEA, 44p.

BEZOT P. 1963. L'amëlioration des sorghos au Tchad. Agronomie Tropicale 18, 985-1007.

CHANTEREAU J. and NICOU R., 1991. *Le sorgho*. Paris, Maisonneuve et Larose, collection le technicien de l'agriculture tropicale, 159p.

CHAVAN U. D., PATIL J. V., SHINDE M. S., 2009. An assessment of sweet sorghum cultivars for ethanol production. *Sugar Tech.* 11 (4): 319-323.

CLERGET B., 2004. *The role of photoperiodism in the yield elaboration of three sorghum varieties grown in West Africa*. Thëse de doctorat de l'INA-PG, PARIS /France. 192 p.

CLERGET B., 2008a. Variabil^ of dëvelopment rate in сиШуё sorghum (*Sorghum bicolor* (L.) Moench) and relationship with photopëriodism. *Cahiers Agriculture 17*, 101-106.

CLERGET B, DINGKUHN M, GOZE E, RATTUNDE HFW, NEY B. 2008b. Variability of phyllochron, plastochron and rate of increase in height in photoperiod-sensitive *Sorghum* varieties. *Annals of Botany101*, 579-594.

COCHEME J. & FRANQUIN P., 1967. Technical report on a study of the agroclimatology of the semiarid area south of the Sahara in West Africa.

CURTIS D. L., 1968. The relation between yield and date of heading of Nigerian sorghums. *Experimental Agriculture,* **4**, 93-101.

DE WET J.M.J., HARLAN J.R. and PRICE E.G., 1970. Origin of variability in the Spontanea complex of Sorghum bicolor. *American Journal of Botany* 57(6) : 704-707.

DOGGETT H., 1965. Striga hermonthica on sorghum in East Africa. *J. Agric. Sci.* 65: 183 194.

DOGGETT H., 1988. Sorghum. London Harlow (GB), Longman Scientific Technical, (2nd ëdition), 512 pp.

DUKE, 1965. "Keys for the identification of seedlings of some prominent woody species in 8 forest types in Puerto Rico," Ann. Missouri. Bot. Gard, n°52, p. 314-350.

EPA/DGESS/MAAH, 2018. Tableau de bord statistique de l'agriculture.

EPSO, Fascination of Plants Day, May 2015.

FAOSTAT, 2014. The State of Food and Agriculture.

FAO, 2002. Sugar sorghum in China. *World Food Summit*, 3p.

GAPILI N., 2016. Etude de la diversite gënëtique et du photopëriodisme d'accessions de sorgho grains [*Sorghum bicolor* (L.) Moench] du BURKINA FASO. Thëse de Doctorat. Université Ouaga I Professeur Joseph KI-ZERBO.

GAPILI N., SAWADOGO M., NANEMA K. R., NEBIE B., SAWADOGO N. and ZONGO J. D., 2015. Study of photoperiodism of sorghum ecotypes in Burkina Faso. Institut Tchadien de Recherche Agronomique pour le Developpement (ITRAD), BP 5400, N'Djamena, Chad. Equipe Genetique et Amelioration des Plantes, University of Ouagadougou 03 BP 7021 Ouagadougou 03, Burkina Faso. ICRISAT-Bamako, BP320 Bamako, Mali. *Copyright © 2015 ISSR Journals*.

GARNER W. & ALLARD H., 1923. Further studies in photoperiodism: the response of the plant to relative length of day and night, US Government Printing Office.

GAUFICHON L., PRIOUL J.L. and BACHELIER B., 2010, What are the prospects for genetic improvement of drought-tolerant crops? FARM, 61 p.

GNANSOUNOU E., DAURIAT A., WYMAN C. E., 2004. Refining sweet sorghum to ethanol and sugar: economic trade-offs in the context of North China. *Bioresource Technology*, 96, 9851002.

GRASSI G., 2001. Sweet sorghum: One of the best world food-feed-energy crop. *Latin America Thematic Network on Bioenergy*, 3p.

GUTJAHR S., 2012. Analysis of morphogenetic and biochemical ctiracteres of interest for the development of dual-purpose grain-bioalcohol sorghum sugars. Thëse de doctorat, universite Montpellier 2, 112 p.

HARLAN J. R. and DE WET J. M. J., 1972. A simplified classification of cultivated sorghums. *Crop Science*, vol. 9, n° 2,172-176.

HOUSE L. R., 1987, *Manuel pour la sélection du sorgho* (2eme edit.). Ed. ICRISAT-Patancheru, 229p.

ICRISAT, 2006 a. Biofuel: energy for the poor. *Science with a human face,* India, 4p.

ICRISAT, 2006b. Sweet sorghum: food, feed, fodder and fuel crop. 18p.

KOROTIMI Thera, 2017. Analysis of gënëtic determinants controlling stem production and composition in sorghum (*Sorghum bicolor [L.]Moench*). Integration of bi- and multi-parental approaches.

KOURESSY M., TRAORE SEYDOU B., 2006. Le photoperiodisme des sorghos africains, une reponse a la variabilite pluviometrique. In : AMMA CIRAD (Ed.), Monitoring and forecasting the African monsoon impacts on agriculture and vegetation (AMMA) Dakar, Senegal. *CIRAD - Agritrop*, 557-559.

KOURESSY M., DINGKUHN M., VAKSMANN M. & HEINEMANN A. B., 2008. Adaptation to diverse semi-arid environments of sorghum genotypes having different plant type and sensitivity to photoperiod. *Agricultural and Forest Meteorology,* 148, 357-371.

KOURESSY M., TRAORE S., VAKSMANN M., GRUM M., MAIKANO I., SOUMARE M., TRAORE P.S., BAZILE D., DINGKUHN M., &SIDIBE A., 2008. Adaptation des sorghos du Mali a la variabilite climatique, *Cahiers Agricultures* vol. 17, no. 2, March-April 2008.

MAISONNEUVE ET LAROSE, sorghum, 1991.

MINISTERE DE L'AGRICULTURE ET DE LA SECURITE ALIMENTAIRE- Situation de reference des principales i'iiieres agricoles au Burkina Faso, Avril 2013.

MURDOCK G., 1959. Staple subsistence crops of Africa. *Geographical Review* 50: 521 - 540.

NAMOANO B., 2018. Effect of sowing dates on agromorphological paramëtres of twelve (12) accessions of sweet grain sorghum [*Sorghum bicolor* (L.) Moench] from Burkina Faso. Memoire de Master. Joseph KI-ZERBO University.

NEBIE B., 2009. Etude de la variabilite agromorphologique de quelques ecotypes de sorghos sucres [*sorghum bicolor* (l.) Moench] du Burkina Faso. DEA dissertation. University of Ouagadougou.

NEBIE B., NANEMA K. R., BATIONO/KANDO P., TRAORE R. E., LABEYRIE V., SAWADOGO N., SAWADOGO M. and ZONGO J. D., 2013. Variation in agromorphological characteristics and brix of a collection of sweet stem sorghums from Burkina Faso. Article publié. *2013 International Formulae Group.*

NEBIE B., 2014. Genetic diversity of sweet stem sorghums [*sorghum bicolor* (l.) Moench] from Burkina Faso. Thëse pour obtenir le grade de docteur a l'Universite de Ouagadougou.

NYABYENDA P., 2005. *Les plantes cultivees en regions tropicales d'altitudes dAfrique.* Belgium, Les Presses Agronomiques de Gembloux. Ed. TEC & DOC, 500p.

OUEDRAOGO M., 2014. Study of the agromorphological diversity of sorghum and identification of cultivars tolerant to post-flower water stress. CAP/Matourkou - Ingënieur d'Agriculture.

QAZI H. A., PARANJPE S., BHARGAVA S., 2012. Stem sugar accumulation in sweet sorghum - Activity and expression of sucrose metabolizing enzymes and sucrose transporters. *Journal of Plant Physiology,* 169 : 605-613.

SAWADOGO B., 2013. Etude de la variabilite morphologique et biochimique de sorghos a tiges sucrees [*Sorghum bicolor* (L) Moench] de la zone soudanienne du Burkina Faso. Memoire de fin de cycle en vue de 1 obtention du diplome d'etudes approfondies (DEA).

SAWADOGO N., OUEDRAOGO M. H., TRAORE R. E., NANEMA K. R., KIEBRE Z., BATIONO-KANDO P., NEBIE B., SAWADOGO M., ZONGO J. D., 2017. Effect Of Agromorphological Diversity and Botanical Race on Biochemical Composition in Sweet Grains Sorghum [*Sorghum bicolor* (L.) Moench] of Burkina Faso. J. BioSci. Biotech. 2017, 6(1) : 263269.

SCHAFFER R. E. and GOULEY L. M., 1982. Sorghum as an energy source. *In: Sorghum in the eighties.* L. R. House, L. K. Mughogho and J. M. Peack, ICRISAT, vol. 2, 477-783.

SENE L., 1995. Reponse de la varite de sorgho CE 145-66 a l'alimentation en eau : effets du stress hydrique sur les rendements et la qualité des semences. Memoire de fin d'etudes pour l'obtention du diplôme d'ingenieur des travaux agricoles. Institut Senegalais de Recherches Agricoles (I.S.R.A.). Senegal.

TRAORE S.B., REYNIERS F.N., VAKSMANN M., KONE B., SIDIBE A., YOROTE A., YATTARA K. & KOURESSEY M., 2000. Adaptation to drought of local sorghum ecotypes in Mali. Science et changements planetaires/ Secheresse. Vol. 11, n° 4, 227-237

VAKSMANN M., BOUCHET S., MALLE *Y*, TOGOLA S., TRAORE P. C. S., 2011. Genetic diversity, structure, gene flow and evolutionary relationships within the *Sorghum bicolor* wild- weedy-crop complex in a western African region. *Theor. Appl. Genet*, 16 p.

VAKSMANN M., KOURESSEY M., CHANTEREAU J., BAZILE D., SANGARD F., TOURE A., SANOGO O., DIAWARA G. & DANTE A., 2008. Utilisation de la diversite genetique des sorghos locaux du Mali, *Cahiers d'études et de recherches francophones/ Agricultures, Vol. 17, Num. 2, 1405.*

VAKSMANN M., TRAORE S. & NIANGADO O., 1996. Photoperiodism of African sorghum. *Agriculture et developpement,* **9,** 13-18.

VIGUIER P., 1947. Sorghum and its cultivation in the French Sudan. Dakar, 80p.

ZOLIKPO S., 2011. Caracterisation agro-morphologique des cultivars traditionnels de sorgho colorant (Sorghum bicolor Sorghum bicolor) au Benin. Universite d'Abomey-Calavi (Benin) - Diplome d'ingenieur agronome 2011.

APPENDIX

APPENDIX: List of sweet stem sorghum genotypes studied (NEBIE, 2014)

APPENDIX 1:

anal number	2	70	73	79	81	85	88	89	92	97
Num coll	BKB5	NBO1	NBO4	BKO2	BZI3	BSA5	BBO5	LTI5	SDJ2	SPO1
dim zone	nord souda	sub sah	sub sah	sub sah	sub sah	sub sah	sub sah	sub sah	sah	sah
Province	Bazega	Namentenga	Namentenga	Bam	Bam	Bam	Bam	Loroum	Soum	Soum
Village	Wemtinga	Boulsa-Sector 4	Boulsa-Sector 1	Darigma	Minimum	Foursa	Boulonga	Titao-Sector 4	Borguindd	PoЬё-Mengao
Local name		kankansiido	kankansiido	loumdi	kankansiido	zouanga	Loumdi	kankansiido	капогё	kendamoongo
Meaning		cane sucree	cane sucree	compact	sugar cane	blind	compact	sweet stem	sugar cane	sorghum sugar

Sah = Sahdlienne

Sub sah = sub-Sahdhian

Nordda = northern Sudanian

APPENDIX 2:

anal number	98	103	104	105	108	109	110	112	115	119
Num coll	SPO3	SAR7	KGA1	KGA2	KGA6	KGA7	KGA8	KBA2	KBA5	KBA9
dim zone	sah	sah	Komondjari	sub sah	sub sah	sub sah	sub sah	sub sah	sub sah	sub sah
Province	Soum	Soum	Оиё	Komondjari	Komondjari	Komondjari	Komondjari	Komondjari	Komondjari	Komondjari
Village	РоЬё-Mengao	Kattd	Gourmantchd	Оиё	Lonadeni	Kokogou	dyapwargou	Bontegou	Bontegou	Bontegou
Local name	kenda	A kaa na	compact	kourbouli-kadëma	kadayoutchala kadayiboana		kankansiido kaDafingrmonga		kadadibimomma kadamengou	
Meaning	sorghum	sweet stem		sweet stem sorghum	sugar cane гесоигЬёе	cane a GLUME SUGAR black	sugar cane	red strawberry	red seeds	sweet stem

Sah = Sahdlienne

Sub sah = sub-Sahdhian

APPENDIX 3:

anal numbe r	120	123	124	125	127	129	130	136	137
Num coll	KBA10	KBA14	GBO2	GBO4	GBO6	GBO8	GBO9	GBI3	GBI1
dim zone	sub sah	sub sah	sub sah	sub sah	sub sah	sub sah	sub sah	sub sah	sub sah
Provin ce	Komondjari	Komondj ari	Gnagn a	Gnagn a	Gnagna	Gnagna	Gnagna	Gnagna	Gnagna
Villag e	Bontegou	Bontego u	Bogan dd- Sector 4	Bogan dd- Sector 4	Bogand d- Sector 6	Thiery	Gnimpie ndi	Bilanga	Hatdry
Local name	kadayitionti onga	fiaribong a	kadem a	hmagli	kankade ma	kahourla iga	kahourla iga	kadayoutc hala	bddipou ba youbili
Meani ng	curved panicle	black sorghum	sugar cane	compa ct	sugar cane	can't see the sun	can't see the sun	curved sugar cane	braids from the chef's wife

Sah = Sahelian

Sub sah = sub-sahehian

More Books!

yes

I want morebooks!

Buy your books fast and straightforward online - at one of world's fastest growing online book stores! Environmentally sound due to Print-on-Demand technologies.

Buy your books online at
www.morebooks.shop

Kaufen Sie Ihre Bücher schnell und unkompliziert online – auf einer der am schnellsten wachsenden Buchhandelsplattformen weltweit! Dank Print-On-Demand umwelt- und ressourcenschonend produzi ert.

Bücher schneller online kaufen
www.morebooks.shop

info@omniscriptum.com
www.omniscriptum.com

OMNIScriptum

MIX
Papier aus verantwortungsvollen Quellen
Paper from responsible sources
FSC® C105338

Printed by Books on Demand GmbH, Norderstedt / Germany